I0066950

Clarence Moores Weed

Stories of Insect Life

Clarence Moores Weed

Stories of Insect Life

ISBN/EAN: 9783744750882

Printed in Europe, USA, Canada, Australia, Japan

Cover: Foto ©berggeist007 / pixelio.de

More available books at **www.hansebooks.com**

STORIES

OF

INSECT LIFE

BY

CLARENCE MOORES WEED

———•—•———

BOSTON, U.S.A., AND LONDON
GINN & COMPANY, PUBLISHERS
The Athenæum Press
1900

69211978

Copyright, 1897, by

CLARENCE MOORES WEED

ALL RIGHTS RESERVED

W42
ENTO

CONTENTS.

		PAGE
I.	THE TENT CATERPILLARS AND THEIR NESTS	1
II.	THE MOTH AND ITS EGGS	6
III.	THE TENT CATERPILLAR PARASITE	10
IV.	THE WORMS ON THE CABBAGE LEAVES	12
V.	THE PARASITES OF THE CABBAGE WORMS	16
VI.	THE RED AND BLACK TIGER CATERPILLAR	18
VII.	THE PARASITE OF THE TIGER CATERPILLAR	20
VIII.	THE MOURNING CLOAK OR ANTIOPA BUTTERFLY	22
IX.	THE CLOUDED SULPHUR BUTTERFLY	25
X.	AN AMBUSHED BANDIT	28
XI.	THE LIFE-HISTORY OF THE ANT-LION	31
XII.	THE MAY BEETLES, OR JUNE BUGS	33
XIII.	THE COMMON POTATO BEETLE	35
XIV.	THE QUEER CASES ON THE WILLOW TWIGS	37
XV.	THE HISTORY OF THE DOBSON	41
XVI.	THE DOBSON BECOMES A HELLGRAMITE	44
XVII.	THE APHIS LION	46
XVIII.	THE GOLDEN-EYE, OR LACE-WINGED FLY	48
XIX.	THE WHITE EGG AMIDST THE PLANT LICE	50
XX.	THE FLOWER SPIDER	52

NOTE.

THE original sources of those illustrations in this book which have not been prepared under the direction of the author should be credited as follows :   After C. V. Riley, Figs. 2, 5, 9–11, 15, 17–22, 33, 34, 36, 39, 41–44, 50; after Brehm, Fig. 12; after Harris, Figs. 23–25; after Forbes, Figs. 31, 32; after Packard, Figs. 27, 28, 30.   Fig. 29 is after an English author, and Figs. 45–49 are reproductions of illustrations published by the Division of Entomology of the U. S. Department of Agriculture.   Many of the other figures were drawn for this book by Mr. James Hall.

C. M. W.

STORIES OF INSECT LIFE.

THE TENT CATERPILLARS AND THEIR NESTS.

OF course every child who has taken a walk in the country in spring has seen the caterpillars' nests in the apple and wild cherry trees. No doubt you thought they were not very pretty, and perhaps you shuddered when thinking of the "horrid worms" you knew were in them.

But if you could sit on a big apple limb some day and watch one of the nests close at hand, I think you would find much to interest you. In the morning, some time after sunrise, you would see the "horrid worms" come out of the doors of the tent and march along — mostly in Indian file — in search of breakfast. When they come to a fork in the branch some will go to the right and some to the left, but each will finally stop when it finds a leaf to its liking. It will then feed upon the leaf, biting it on the edges with its good-sized jaws, and often leaving only the midrib to show that a leaf was there.

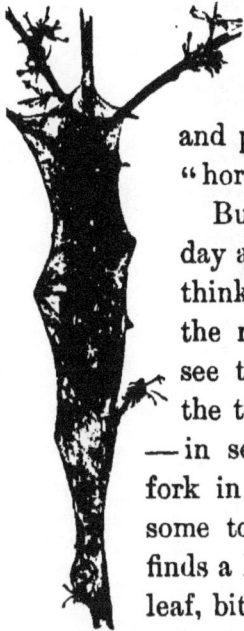

FIG. 1.
Caterpillar's
Nest.

After breakfasting an hour or two, most of the caterpillars are likely to march back to the tent and crawl in through the half-closed doors, where they range themselves side by side, much as sardines are packed in a box. By thus seeking shelter

1

during the middle of the day, they hide away from the birds and
from some little flies that are always looking for caterpillars to
lay their eggs in them.   But I will tell you later why the flies
do this.

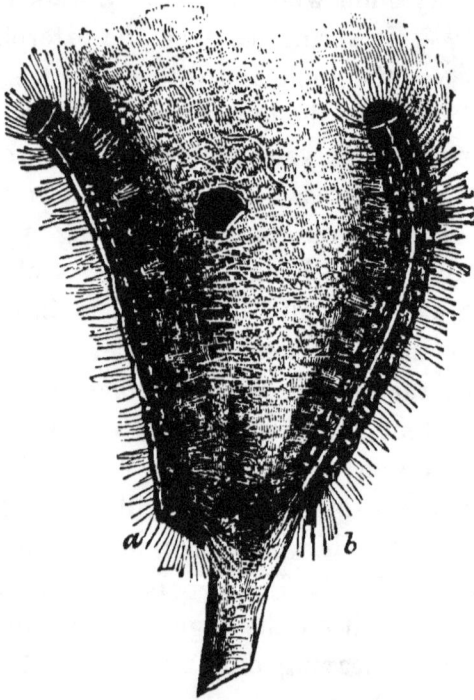

FIG. 2. — Full-grown Tent Caterpillars.

Towards the middle of the afternoon the procession of the
caterpillars may again be seen going forth to war upon the un-
resisting leaves.   As in the morning, they scatter here and
there over the twigs, each choosing a leaf for its victim and
devouring it piecemeal until hunger is satisfied.   Then homeward they go, and through the cold, damp night they keep each
other warm beneath the silken folds of the tent.

If you look carefully at the surface of the limb on which the procession has been marching you will see many whitish silken threads.   One of these threads is spun by each of the caterpillars as it marches along.   The thread comes from the

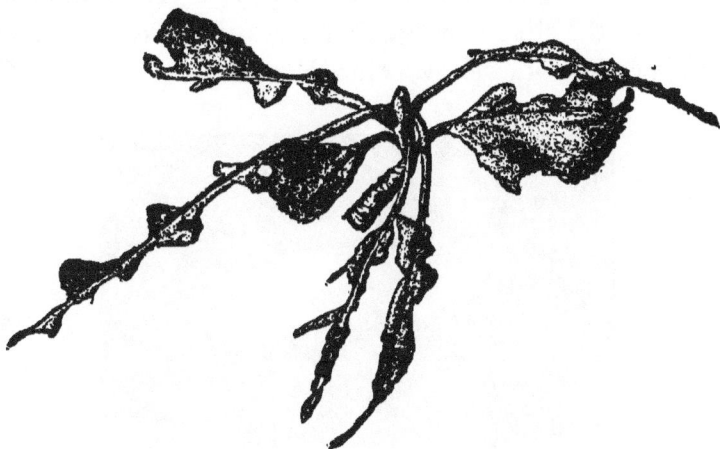

FIG. 3. — Apple Leaves eaten by Tent Caterpillars.

mouth in the form of a liquid, secreted by certain peculiar glands, which on exposure to the air hardens into a silken thread.   Probably the caterpillar is guided back to its tent by the thread which it spun on the outward journey.

During cold and wet weather the caterpillars remain within the shelter of the tent, sallying forth again when the spring sun shows his genial face.

About the first of June the caterpillars become full-grown, as far as this part of their life is concerned.   They can eat no more leaves, and appear to be seized with a desire to wander away from home.   Down the tree they crawl and out into a strange new world — a jungle of weeds and grass — they go, seeking here and there the friendly shelter of a stone or board or fence.   When such shelter is found, the caterpillar halts for rest. Soon it begins weaving about itself a silken shroud, the glands in the mouth which furnish the thread to guide it homeward from its feeding grounds again doing duty for the shroud. Before long the caterpillar is hidden within the white silken woof. It next ejects from its body a yellow fluid which runs among the silken meshes and gives the *cocoon* — for so the shroud is called — a yellow color (Fig. 4).

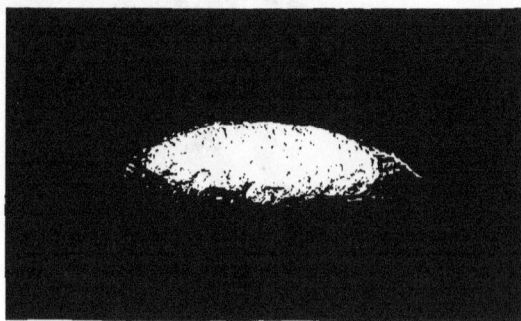

FIG. 4. — Cocoon of Tent Caterpillar.

The body of the caterpillar now becomes shorter and thicker. Before long the skin on the front part of its back splits open and the caterpillar wriggles violently until the skin is finally crowded off to the hinder end, and there lies within the cocoon only a brown chrysalis.

The change from the active caterpillar to the quiet chrysalis

FIG. 5. — Moth of Tent Caterpillar.

is a strange *transformation*. The chrysalis takes no food, and its only movement is a feeble wriggle. The insect remains in this condition for nearly two weeks. Then another change takes place: the skin of the chrysalis splits apart and there comes forth a queer-looking moth that pushes its way through the meshes of the cocoon. When its wings are finally spread out and dried it resembles Fig. 5, if it happens to be a female moth. If it is a male moth it is somewhat smaller. The color in both is reddish brown.

Thus the caterpillar has reached the highest stage of its existence. Within a few weeks the "horrid worm" has become a handsome moth.

THE MOTH AND ITS EGGS.

IF you could become a fairy small enough to ride upon the back of one of the larger of these reddish-brown moths, you would have an interesting experience. During the day the moth would hide with you in almost any quiet shelter she could find, but at night she would fly abroad with many other moths of the same and other kinds. She might be attracted by the light shining through somebody's window, and bump your fairy nose against the pane. But more likely she would ask you to rest upon an apple twig while she busied herself in laying her eggs. She fastens these upon the twig in clusters of two hundred or more, setting them on end side by side upon the bark. When the laying of a cluster is finished the moth covers the eggs with a glue-like substance, which hardens into a shiny varnish that keeps out the moisture (Fig. 6).

FIG. 6. — Egg Mass of Tent Caterpillar.

After the eggs are laid the fairy will do well to find another moth to carry its tiny self, for this moth will soon die, her purpose in life being accomplished when the eggs were laid.

These eggs fastened upon the twigs of apple and wild cherry trees during July do not hatch until the following spring. The marvelous change within the shell by which the egg develops into a tiny caterpillar takes place, however, before winter begins. If you could carefully open one of the little cylinder-like eggshells during cold weather you would find the fully formed caterpillar within. It is such a condition as would occur if a hen's egg developed into a chick which remained alive inside the shell for several months before pecking its way out.

When the long months of waiting through the cold winter are passed, the spring sunshine wakens the caterpillars to life. Then they gnaw through the thin eggshells and crawl out to find themselves in a strange new world. Beside them are the buds bursting into leaf, and, led by that strange knowledge which we call instinct, the band of little caterpillars crawls down the twig to the nearest fork in the branches. Here they spin a silken web which is the beginning of the tent or "nest." They stay in it at night and at other times when not feeding upon the leaves.

About a week after the caterpillars have hatched, their bodies have so increased in size that they must provide themselves with a skin larger than the one with which they were born ; for insects do not grow as the higher animals do. With the latter the skin grows along with the body, but with the former it does not stretch and cannot increase in size. So some day the colony of caterpillars remains at home beneath the silken folds of the tent. The skin of each splits open along the back, and the caterpillar crawls out of the old skin clothed in a new one that had been formed beneath the other.

When the caterpillars become used to the new clothes thus so kindly provided by Mother Nature, they sally forth again in search of food. This skin-shedding process is called *moulting*. It is repeated several times during the lives of the caterpillars, which become full-grown in about six weeks. They then resemble Fig. 2, and are nearly two inches long. The body is hairy and has a distinct white stripe along the middle of the back, on each side of which are short yellow lines. The sides are partially covered with paler lines, spotted and streaked with blue. The lower surface of the body is black.

FIG. 7. — Wild Cherry Tree, showing Nests of Tent Caterpillars.

You can easily watch the growth of these caterpillars by placing two or three of them in a glass-covered box, having a little sand or earth in the bottom. Feed them every day with apple or wild cherry leaves freshly dipped in water.

In feeding, the tent caterpillars devour the substance of the leaf, often taking all but the midrib, but more commonly leaving parts of the leaf along the midrib as shown in Fig. 3. If many caterpillars are on the tree, its leaves will be entirely eaten off, a condition shown in the photograph reproduced in Fig. 7.

THE TENT CATERPILLAR PARASITE.

IN many of the nests of the tent caterpillars you can find peculiarly shrunken caterpillar skins, looking like Fig. 8. The under side is generally split open and shows part of a silken cocoon.

FIG. 8. — Tent Caterpillar killed by a Parasite.

These are the remains of caterpillars which have been killed by parasites. Weeks before, when the caterpillars were rather small, an egg was deposited in each by a four-winged fly, resembling in general appearance Fig. 11. The egg hatched into a tiny maggot, which grew by absorbing the juices of the caterpillar's body.

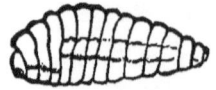

FIG. 9. — Parasitic Maggot.

FIG. 10. — Pupa of Parasite.

In two or three weeks the maggot became so large that the caterpillar was killed and nothing was left of it but the skin with the parasite on the inside. The latter then spun a silken cocoon, within which it changed to a pupa (Fig. 10).

A short time afterwards another change takes place and from the pupa there emerges a fly similar to the one which laid the egg in the young caterpillar (Fig. 11). This fly is called an ichneumon fly.

FIG. 11. — Ichneumon Fly.

The tent caterpillars have many enemies besides these ichneumon flies. While their hairy skins protect them from the attacks of many birds, there

10

are some, like the Blue Jay and the Cuckoos, which devour them eagerly. In one case a Cuckoo was seen to eat twenty-seven caterpillars, one after the other, at one time, all taken from a single nest.

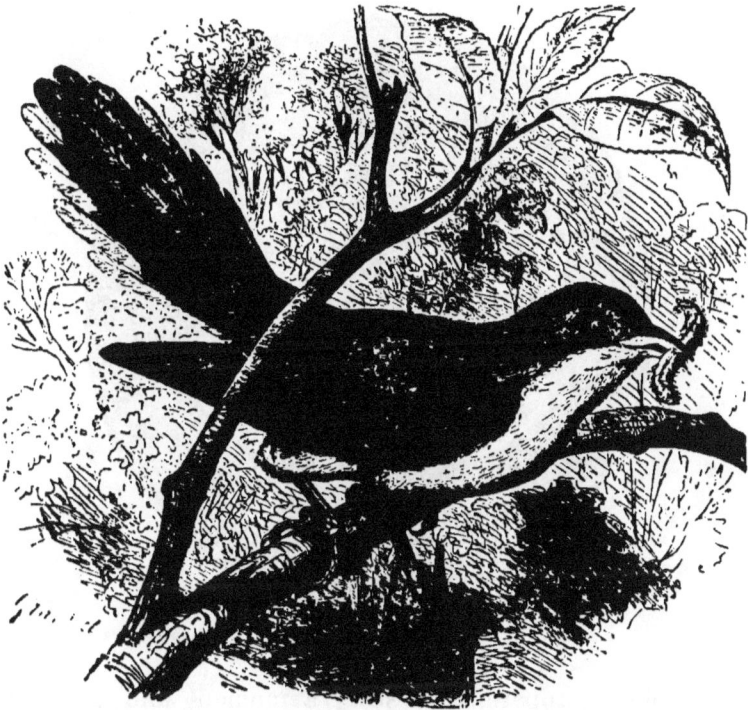

FIG. 12. — Yellow-billed Cuckoo.

EVERY boy has seen the pretty white butterflies represented in Figs. 13 and 14, flying about in spring and summer. Now one will stop to suck the nectar from a clover blossom, while another sips the sweets from a buttercup. Again you will see them in the garden, lighting here and there upon the young

FIG. 13. — White Cabbage Butterfly. Male.

cabbage plants, apparently resting a moment and then flying away.

If you pick off a leaf upon which such a butterfly has rested, you will be likely to find a tiny whitish object of a somewhat conical shape firmly fastened to the surface. This is the egg of the butterfly.

12

If you remove the cabbage leaf to which the egg is attached, and place it in a covered tumbler, or a box with a piece of glass laid over the top, you will be able by careful attention to learn many interesting things about the life-history of the insect.

A few days after the egg is laid it will be likely to hatch into a tiny, green, worm-like caterpillar, that will nibble at a

FIG. 14. — White Cabbage Butterfly. Female.

fresh piece of cabbage leaf which you should furnish. It will continue to eat in this way, from day to day, for about a week. Then if you watch it closely you may be able to see it undergo the curious process of shedding its skin, called *moulting*. This is the way in which caterpillars provide for their increase in size.

Before moulting begins the little cabbage worm rests from eating. Then its skin splits open along the back, especially toward the front end of the body, and the worm crawls out, clothed in a new skin which has been formed beneath the old one. It soon begins to eat the cabbage leaf again.

The caterpillar is larger after this *moult* than before. It feeds quietly for another week, and then it sheds its skin again, being considerably larger after the second moult. Thus the processes of eating and moulting are continued for perhaps two weeks longer. Then the cabbage worm becomes full-grown, as far as this caterpillar or *larva* stage of its life is concerned (Fig. 15, *a*).

The larva is now ready to enter upon the next stage of its existence — that of the *pupa* or *chrysalis*. The caterpillar seeks the side of its cage, or if out-of-doors it finds some secure shelter, and spins a silken loop to hold up the front part of its body. It also spins a tuft of silk in which to entangle its hind legs. Meanwhile its body becomes short and thick. Then along the back a crack appears, and the pupa wriggles out of the larva skin. The latter falls off, leaving only the chrysalis attached to the tuft of silken threads at one end, and held by the loop at the other end (Fig. 15, *b*).

FIG. 15. — *a*, Cabbage Worm; *b*, Chrysalis.

The insect remains in this chrysalis state about two weeks. Then the skin of the chrysalis splits open and a white butterfly comes out of it. At first its wings are small and cramped, but they soon spread out and the butterfly takes wing, if given its liberty.

This butterfly is much like the one which laid the egg at the beginning. It has thus gone through the four great stages of insect life. It existed first as an *egg*, then as a *larva*, then as a *chrysalis*, and finally as an *adult* insect or butterfly. These various changes are commonly spoken of as the *transformations* or life changes of the insect.

In appearance the two sexes of the butterflies differ slightly from each other. In the males (Fig. 13) there is but one black spot on each front wing, while in the females (Fig. 14) there are two.

To study the life-history of this insect it is not necessary to raise it from the egg. Worms of various sizes may be obtained from cabbage plants and reared in glass-covered boxes. They should be frequently provided with fresh cabbage leaves. The later stages of the insect's life-history may thus readily be observed.

THE PARASITES OF THE CABBAGE WORMS.

I SHOULD not be surprised if some morning in the box in which you were rearing the cabbage worms collected out-of-doors when partly grown, you found only a shapeless withered skin in place of a fat green worm. And beside the skin you will see twenty or thirty small yellow bodies, lying in an irregular mass (Fig. 16). Each of the bodies, on closer view, you will see are composed of fine silken threads.

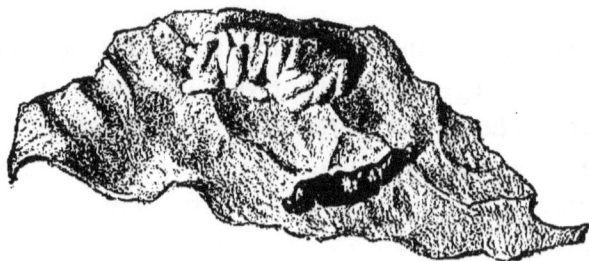

FIG. 16. — Cocoons of Cabbage Worm Parasite.

These are the *cocoons* of another kind of insect. The way they came where they are is explained in this way: some weeks before, when the cabbage worm was only partially grown, a small, black, four-winged fly (resembling Fig. 11) lit upon the worm while feeding or resting on the cabbage plant, and deposited beneath its skin twenty or thirty tiny eggs.

In a few days these eggs inside the cabbage worm hatched into minute footless maggots, which remained within the body of the caterpillar, absorbing its life-blood. On this account the maggots are called *parasites*. In order to grow they rob another creature of its life.

The parasitic maggots finally become full-grown in this larval stage of their life. Then, all acting at the same time, they

16

work their way out of the skin of their *host*, as the unfortunate cabbage worm in which they develop is called. After the parasites have emerged, the caterpillar becomes merely a dead and shapeless mass of skin.

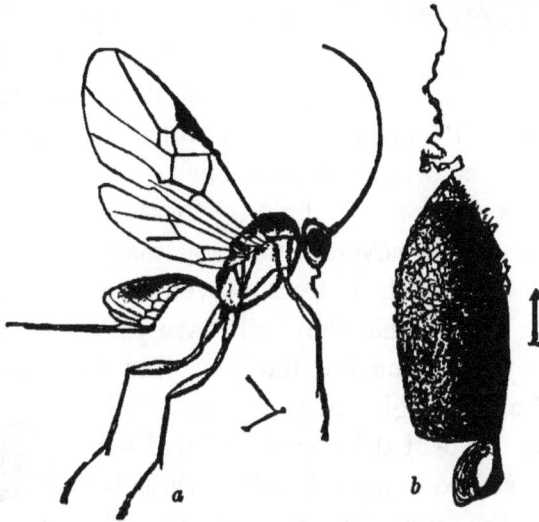

FIG. 17. — Parasitic Fly and its Cocoon, much magnified.

The maggots so strangely free from the caterpillar whose life-blood they have absorbed crowd together on the leaf. Each spins by means of a liquid silk a sheltering cocoon. In these cocoons the maggots change to pupæ.

If such a mass of cocoons is placed in a bottle, or a covered tumbler, a week or two later you are likely to find in the vessel a number of small black flies. Each cocoon will have a tiny cap partly open at one end, showing where the flies have come out.

These flies are the parasites full-grown. They are similar to the fly which laid the egg in the cabbage worm some weeks before. Set them at liberty and they will go forth in search of other cabbage worms, to repeat the tragic cycle of insect life.

THE RED AND BLACK TIGER CATERPILLAR.

EVERY boy must be familiar with the red and black hairy caterpillar represented in Fig. 18. It is one of the commonest insects in spring and autumn, and is often seen even in winter.

FIG. 18. — Tiger Caterpillar.

This insect is so abundant in many regions that it is often spoken of as *the* caterpillar. People think of it rather than of other kinds when the word caterpillar is mentioned.

If you pick up two or three of these caterpillars in spring and place them in a covered box, you may readily study their life changes. Give them fresh clover leaves dipped in water every day or two, so that they will have plenty to eat.

Before very long you are likely to find in place of one of the caterpillars a rough, hairy cocoon. This appears to be made of the cast-off hairs of the caterpillar, fastened in place by silken threads. Inside of this the caterpillar becomes a smooth brown pupa or chrysalis (Fig. 19).

FIG. 19. — Cocoon and Chrysalis.

A week or two after the cocoon was formed the chrysalis splits open and a brown moth emerges, pushing its way through one end of the cocoon. It is of medium size. When its wings expand it resembles Fig. 20. It is yellowish brown with spots on the wings.

The moths fly about at night. They frequently enter lighted rooms through open windows. They are called the isabella tiger moths.

The caterpillars of the isabella moth feed upon many kinds of plants. But they seldom injure to a serious extent any

18

cultivated crop. This is probably on account of their solitary feeding habits and the numerous parasites that prey upon them.

Prof. J. H. Comstock has published the following graphic account of the habits of these caterpillars : " ' Hurrying along like a caterpillar in the fall' is a common saying among the country people in New England, and probably had its origin in observations made upon the larva of the isabella tiger moth. This is the evenly clipped, furry caterpillar, reddish brown in the middle and black at either end, which is seen so commonly in the autumn and early spring. Its evident haste to get somewhere, in the autumn, is almost painful to witness. A nervous anxiety is evident in every undulating movement of its body, and fre-

FIG. 20. — Isabella Tiger Moth.

quently its shining black head is raised high in the air and moved from side to side while it gets its bearings. Occasionally after such an observation it evidently finds it is mistaken and turns sharply and hastens along faster than ever in another direction. So far as we can judge, its excitement comes from a sudden fear that winter will overtake it before it can find a cosy protected corner in which to pass its winter sleep."

The hairy caterpillars of this and some closely related moths are often called " woolly bears " in New England. In other regions this species is named " the hedgehog caterpillar " on account of its habit of curling up when disturbed.

THE PARASITE OF THE TIGER CATERPILLAR.

DURING the spring months the brown cocoons of the tiger caterpillar may be found under boards, logs, fence-rails, and similar shelter beside fields and highways. It will be worth your while to collect a dozen or more of them and keep them indoors in a glass-covered box.

During the next two or three weeks the brown moths are likely to emerge from the cocoons, but from a few of the cocoons there are likely to come forth one or more entirely different creatures. These are slender bodied insects, having four transparent wings. They resemble bees or wasps in their general appearance.

FIG. 21. — Ichneumon Fly.

These peculiar insects are ichneumon flies. The way they came to be in the cocoon of the tiger caterpillar may be explained thus : Many weeks before, a four-winged fly laid one or more eggs in the body of the tiger caterpillar. The egg soon hatched into a tiny whitish maggot which absorbed the blood and other liquids in the body of the caterpillar. It continued to grow for some time, and at last, after the caterpillar had spun its cocoon, the maggot killed the

FIG. 22. — Larva of Ichneumon Fly.

caterpillar and changed to a pupa inside the cocoon. There it remained until the time for another change. Then it became a full-grown ichneumon fly and gnawed a hole in the cocoon.

It crawled out through the hole and was ready to fly away. In a short time, if out-of-doors, it would search for other caterpillars in order to lay eggs in them.

You will see from this account of the life of the ichneumon fly that it is a *parasite*. There are many different kinds or *species* of these ichneumon parasites. One of the largest of those which attack the tiger caterpillar is represented natural size in Fig. 21.

THE MOURNING CLOAK OR ANTIOPA BUTTERFLY.

BUTTERFLIES in winter are rare objects, but by a little searching in the right situations you may often find specimens of the beautiful mourning cloak or antiopa butterfly. This is the large purple butterfly having a golden border on the upper surface of the wings, shown in Fig. 23.

FIG. 23. — Mourning Cloak or Antiopa Butterfly.

On the approach of winter these butterflies seek the shelter of logs, boards, hollow trees, wood piles, culverts or bridges, sheds or barns. When one of them finds a shelter to its liking it lights, generally with its wings hanging downward. Here it composes itself for the long sleep which knows no waking till the spring sunshine calls it again to life.

In such shelters protected from snow and rain, these fragile creatures endure without injury the severe cold of our northern winters.

When the warm rays of the April sun have penetrated to the retreats of these butterflies, they come forth to flit slowly about in sunny glades, seeking shelter at night and during stormy weather. Their beauty is likely to be somewhat faded, but the early stroller in the woods is gladdened by their sight, as they sail through the air above the lingering drifts of snow, or plunder here and there the nectar from the first willow " pussies."

As spring advances the mourning cloak butterflies visit the fragrant blooms of the arbutus and other flowers.   From these they suck the nectar, and in return for the sweet food they carry the pollen from flower to flower.

When the settled warmth of May has brought out the leaves of the trees and shrubs, these butterflies lay their eggs in clustered rows upon the twigs of elm, willow, and poplar.   Then the butterflies, having lived to what for a butterfly is a green old age, die one by one.

About two weeks after the eggs are laid they hatch into tiny caterpillars.   Each caterpillar gnaws the eggshell around its upper edge until a circular lid is cut out.   This is pushed upward and the caterpillar crawls out.

After thus hatching, all the caterpillars from one cluster of eggs crawl upon a neighboring leaf.   Here they " range themselves side by side in compact rows " with their heads toward the edge of the leaf.   Then they feed together upon the green substance of the leaf, but do not eat the veiny framework.

As the caterpillars grow they seek new leaves.   In doing this they have to scatter more or less because one leaf will not support them all.   But they remain close together upon one twig, the leaves of which suffer sadly.   For the growing caterpillars no longer are content with the green surface, but eat all the leaf except the larger veins.   Before the caterpillars stop eating even the principal veins are devoured, and only the midribs are left.

The caterpillars moult or cast their skins several times during their growth, huddling together on a branch for this purpose, and leaving behind a spiny mass of skins when the process is completed.

The full-grown caterpillar of the mourning cloak butterfly is represented natural size in Fig. 24. Its general color is dull black ; the head is more or less tinged with bronze. Beginning with the third ring behind the head there is along the

FIG. 24. — Caterpillar of Antiopa Butterfly.

middle of the back a row of good-sized orange red spots. The larger legs are of a somewhat similar color. Along each side of the back are rows of blackish barbed spines, giving the caterpillar a formidable appearance.

These caterpillars become full-grown in June. Each then seeks some shelter where it changes to a chrysalis of a yellowish brown color, about an inch long (Fig. 25). It remains in this condition about two weeks. Then it comes forth as a butterfly.

FIG. 25. — Chrysalis of Antiopa Butterfly.

These midsummer butterflies soon deposit their eggs for another brood of caterpillars which go through their changes and become butterflies in September. It is this brood of mourning cloaks that pass the winter in sheltered places.

THE CLOUDED SULPHUR BUTTERFLY.

TAKING the season through, probably no butterfly is more familiar than the common sulphur yellow species, frequently mentioned in books as the philodice butterfly (Fig. 26). It is also often spoken of as *the* yellow butterfly, and occasionally is called the clover butterfly. Mr. Samuel H. Scudder, the eminent student of butterflies, has adopted for it the name of clouded sulphur butterfly, which is a very fitting one.

FIG. 26. — Clouded Sulphur Butterfly.

It is a comparatively simple matter to follow the life-history of the clouded sulphur butterfly through a cycle of its existence. Place a clover stem in a bottle of water, with the leaves projecting upward, and put it under some such shelter as is furnished by an open-mouthed bell-jar or a large paste-

board box. Out-of-doors catch three or four of the yellow
butterflies, and place them in the vivarium thus made, covering the open top with a pane of glass, and putting inside a
dish of sweetened water for the butterflies to sip. In a few
days you are likely to find upon the upper surface of many of
the younger clover leaves one or more small yellowish white
eggs, which become brighter colored as they grow older.

Now remove the butterflies and watch the eggs daily. In
three or four days from the time they were laid they become of
a reddish orange color ; and in five or six days they hatch into
tiny yellowish or brownish caterpillars that will nibble holes in
the young clover leaves. In two or three days each little
caterpillar will cast its skin and become grass green in color.

The caterpillars continue to feed and grow for two or three
weeks, retaining the green color, although they moult three
times after the first moult already mentioned. They feed freely
upon fresh leaves of red or white clover, but in confinement
are likely to cause more or less trouble from their habit of
dropping from the plant upon the least disturbance, — a habit
which doubtless is of the greatest value to them under natural
conditions, because it affords so easily a means of escaping from
enemies of many kinds.

" As the time approaches for the change to the chrysalis,"
writes Mr. W. H. Edwards, " the larva seeks the protection of
some stem, bit of bark, or fence rail, spins a button of pink
silk and a light web over the surface of the object, fastens its
hind feet in the one and its fore feet in the other, and hangs
with its back curved downward or outward. Gradually the
markings of the body become obliterated, lost in uniform green.
In this condition the larva rests for some hours, then rousing
itself spins a loop of several threads from the base of the feet
on one side to a like point on the other, instinctively knowing

just how to make the threads ; and presently, seizing the loop in its jaws, manages to throw it partly over the head, and by a great effort works it entirely over and down the back to the fourth segment, and stops exhausted. Some hours pass without any motion, when suddenly the skin splits on the back of the anterior segments and is rapidly shuffled off, exposing the chrysalis which rests secure on its girdle of silk." About ten days later the butterfly emerges. Except in the far north, there appear to be three broods of the butterflies each season.

These butterflies may often be seen during summer gathering in great flocks by muddy pools in the road, where they come to drink. They also visit clover blossoms and many other flowers to sip the nectar through their slender tongues.

EVERY country boy knows that while many insects feed upon the leaves and fruits of plants, there are some which feed upon other insects. Sometimes they roam about in search of prey, and sometimes they lie in wait in cunningly devised traps to ensnare their victims.

Of those having the latter habit perhaps none is more curious than the ant-lion (Fig. 27), an insect which gets its name because it feeds on ants. Should you see one of these ant-lions on a table or other smooth surface, you would think it one of the clumsiest creatures you ever saw. It is so slow and awkward in its movements that one would think it doomed to starve if its living depends on catching ants.

FIG. 27.
Ant-lion.

But if you will put the ant-lion upon the surface of a loose sandy soil you will soon find that it is able to take care of itself. It has a method of its own that enables it to get a living in spite of its clumsiness.

Soon after you put the ant-lion on the ground you will see it begin to dig a hole. " How can it dig," you ask, " without shovel or spade ? " Look and see the flattened head which the creature works beneath some of the sandy soil. Then see the head jerked suddenly upward, so that the sand on top of it is sent flying some distance away. The insect is no mean shoveler, and before long it excavates in this curious way a cone-shaped pit an inch or more in depth and as steep as the sand will lie (Figs. 28 and 29).

When this strange pitfall is finished, the ant-lion buries itself in the sand at the bottom. The body is concealed, except part of the head and the immense jaws, which are spread wide

28

open. What is the creature waiting for? Were you to watch patiently you would see an ant roaming about over the sand

FIG. 28. — Ant-lion and its Trap.

in search of food. Suddenly it comes to the brink of the cone-like trap, the sand gives away beneath its feet, and it falls into

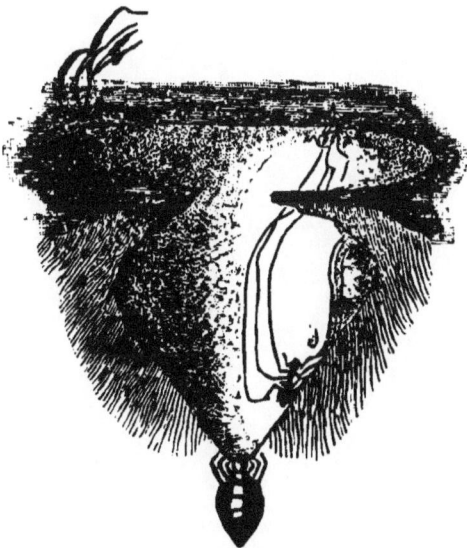

FIG. 29. — Trapping of the Ants.

the jaws of the ambushed bandit at the bottom. There the victim is soon disposed of. It is caught and killed by the jaws projecting from the sand, and its juices are sucked into the stomach of the ant-lion. Afterwards the shrunken skin is thrown out of the pit by the jerking motion of the head.

When it has thus disposed of one victim, the ant-lion returns to its hiding-place to await the coming of another.

Sometimes the sand composing the trap is so damp that the ant after falling part way down recovers its foothold and starts to climb out. Then ensues an exciting contest. The ant-lion tries to prevent the escape by digging away the sand beneath the victim and jerking the particles upwards. It often looks as if the ant-lion tries to hit the escaping prisoner with the sand. Sometimes it succeeds in the attempt.

THE LIFE-HISTORY OF THE ANT-LION.

LIKE most other insects, the ant-lion passes through four distinct stages of existence; namely, (1) the egg; (2) the larva; (3) the pupa or chrysalis; and (4) the adult or imago. The stage we have been discussing is that of the larva, which corresponds to the caterpillar stage of the butterflies.

It is in this larval stage that the insect grows in size. It is supposed that the ant-lion lives as a larva for two years: then it buries itself in the sand and spins around itself a silken case, called the cocoon, which has many particles of sand mixed with its texture. Within this cocoon the ant-lion casts its skin and changes

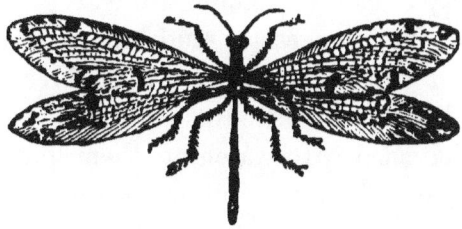

FIG. 30. — Adult Ant-lion.

to the quiet pupa state. It remains in this condition a few weeks; then another change takes place, and there emerges from the cocoon a large four-winged insect (Fig. 30) looking something like our common dragon flies, and very different from the larval ant-lion by which the cocoon was made.

In this winged condition the insect has become an adult and has reached the highest stage of its existence. Its career is now nearly at an end. It has little to do except to deposit eggs in the sand. These eggs will hatch into another brood of ant-lions which will have a similar history.

Ant-lions are common over a large part of the United States. They are most abundant in the south, but occur in many of the northern states. In Illinois and Florida I have found them

31

abundant in the sandy soil beneath overhanging ledges or by the sides of fallen logs.

In many localities these insects are called "doodle bugs," because when boys put their mouths close to the pits and call "doodle, doodle" repeatedly, the little creatures, probably thinking some insect is at hand, come to the surface. Some charming verses concerning this practice have been written by James Whitcomb Riley.

You may easily study the habits of ant-lions by placing them in pans or dishes holding three or four inches of dry sand. Feed them with ants or other wingless insects. When you are satisfied with watching them, put the ant-lions where they can get their own living and do not cruelly leave them to starve.

THE MAY BEETLES OR JUNE BUGS.

DURING the warm evenings of May and June, lighted rooms are frequently invaded by the great clumsy May beetles or June bugs (Fig. 31) which fly through the open doors and windows, being attracted by the light. On such evenings by careful searching out-of-doors one is likely to find these beetles feeding upon the leaves of apple, cherry, oak, elm, hickory, and various other fruit and shade trees. These insects fly and take their food at night, remaining quiet during the day.

FIG. 31. — Adult May Beetle.

The life-history of these May beetles may be briefly summarized in this way: The female beetles deposit their minute whitish eggs among the roots of grasses. The eggs soon hatch into small, brown-headed grubs or larvæ that feed upon the roots of grass and other herbage. They increase slowly in size, burrowing about in the earth to get food, and going down deeper when winter approaches. The next spring they come near the surface again and continue to feed throughout the second season. In autumn they become full-grown and form oval cells in the earth in which they change to pupæ, again to transform, generally before winter, into beetles that remain in

FIG. 32. — White Grub.

33

the cells until the following spring.   They then work their
way out of the ground and fly freely about.

The larvæ of these beetles are the well-known white grubs
or "grub worms" (Fig. 32) so often found in gardens and
plowed lands.   They are large, roundish worms with the body
coiled in the middle, having the head and its organs of a deep
brown color, and three pairs of brown legs attached to the
body just back of the head.

NEARLY every child who has seen potatoes growing in fields or gardens has seen the brown striped beetles, so commonly called "potato bugs," represented in Fig. 33, *d*. Those who have lived upon farms will know that these beetles may be found in the potato field soon after the plants come up, eating the tender leaves, and laying upon the under surfaces of the

FIG. 33. — Transformations of the Colorado Potato Beetle.

latter masses of orange-colored eggs (Fig. 33, *a*). These eggs are sometimes deposited also upon the leaves of grasses, smart-weed, or other plants in the field. A week or more later they hatch into little grubs that feed upon the leaves, gradually increasing in size (*b, b, b*) and occasionally moulting or shedding their skins. In a few weeks they finish their larval growth ; they then descend to the ground, where just beneath the soil surface or under rubbish above it they change to pupæ, emerging as perfect beetles about ten days later. The number of broods varies with the latitude, there being from two to four each season.

This insect was originally a native of the Rocky Mountain region, where it fed upon a wild plant related to the cultivated potato. When the garden patches of the settlers extended to its habitat, so that there were potatoes growing at short distances apart throughout the region between the Rocky Mountains and the Atlantic Ocean, these beetles began feeding upon the new food plant and rapidly spread eastward, until within a few years from the time they started they had reached the eastern states. Then they were carried to various European countries by means of steamboats and sailing vessels.

These potato beetles have a few enemies to contend against. Their eggs are greedily devoured by ladybird beetles and their larvæ, and the other stages are eaten by the Rose-breasted Grosbeak and perhaps a few other birds.

THE QUEER CASES ON THE WILLOW TWIGS.

THROUGH the winter and early spring, a sharp-eyed boy or girl may frequently find on the bare twigs of willow shrubs,

FIG. 34.—Winter Case of Viceroy Caterpillar.

curious cylindrical cases of a somewhat silken texture, attached where a leaf has been. The shape of one of these cases is shown in Fig. 34.

If you examine such a case closely you will find that it is indeed made out of a leaf. You can see the long midrib projecting beyond the round part, and the shape of the leaf stem may be traced below the silky covering. With a little study you will see that the blade of a leaf has been rolled into a tube and its edges sewed together.

If you will carefully open one of these cases you will be likely to find on the inside the cunning creature that has so deftly made a house for itself out of a leaf, that otherwise would long since have fallen to the ground.

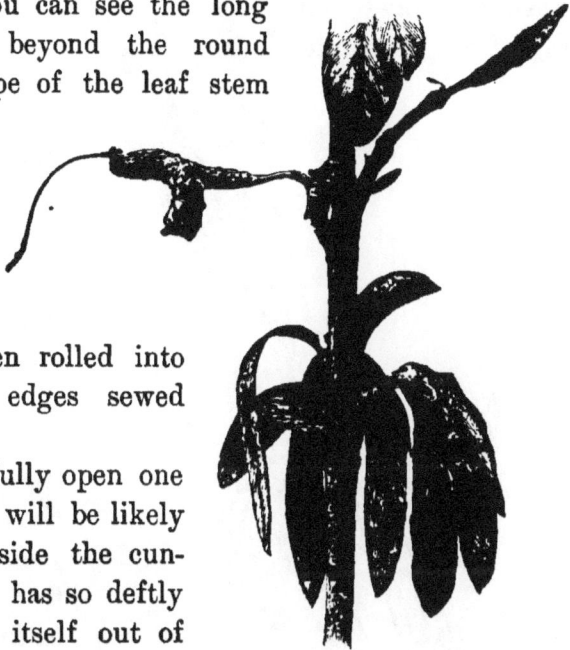

FIG. 35.—Caterpillar resting on Winter Case.

It is a little brownish caterpillar, more or less mottled with

37

black and white, and having many hairs and tubercles on its body.

Should you be able to watch one of these queer willow cases when the "pussies" are appearing you might see the caterpillar, recalled to life by the spring sunshine, crawl out backward from its winter home.

FIG. 36. — Viceroy Caterpillar.

Then it searches for one of the willow "pussies" and nibbles at it to satisfy the hunger brought on by its long fast. It continues to feed upon the "pussies" day after day until the leaves appear. Then it eats them.

The caterpillar is likely to remain near the case in which it has spent the winter, and to return to it when not eating. It then often rests motionless upon it for hours during the day. Its mottled colors render it inconspicuous, or make it look like the droppings of a bird. One of these cater-

FIG. 37. — Chrysalis of Viceroy Butterfly.

pillars thus resting on its case is represented in Fig. 35.

The caterpillars feed most freely after dark, when they are not afraid of being seen by birds. They then leave their resting-places and crawl to the neighboring leaves.

During May the caterpillars become full-grown, when they look like Fig. 36. They then change to mottled brown and gray chrysalids, first fastening themselves to the branches of the willow shrubs.

The chrysalis of the viceroy butterfly is represented in Fig. 37. It is generally a little less than an inch long, and has a peculiar, rounded, wedge-like hump near the middle of the back.

FIG. 38. — Viceroy Butterfly.

During June there emerges from the chrysalis a handsome reddish brown butterfly, streaked and veined with black. This is the viceroy butterfly, represented natural size in Fig. 38.

The viceroy butterflies that have thus appeared in June flit leisurely about in the summer sunshine, sipping the nectar of flowers and searching out willow and poplar shrubs or trees. Having found these, the butterflies deposit their eggs on the tips of the pointed leaves, generally on the upper surface, and usually only one egg on a leaf.

A week later the eggs hatch into tiny caterpillars, that first eat up the eggshells from which they have emerged. Then they feed upon the green surface of the leaves. As they grow older they devour all but the midribs of the leaves attacked.

During their caterpillar growth these insects moult four or five times, eating the cast-off skin each time. Early in July they change to chrysalids, and a little later come forth as butterflies.

You can easily rear any of these caterpillars that you may find, by placing them in a glass-covered box, having a little moist earth in the bottom. Feed them every day with fresh willow leaves dipped in water.

No place is more attractive for a spring excursion than the rock-bound shores of a creek or river. On such an outing a sharp-eyed boy will often see large white patches flattened against the gray rocks hanging over the water. Sometimes single circles will stand out in bold relief, while in other places dozens of white splashes may be crowded together side by side to form irregular masses (Fig. 40).

FIG. 39.—Egg Mass.

FIG. 40. — Egg Masses on a Rock.

If you examine one of these patches (Fig. 39) close at hand, you will find that it is a thin, wafer-like circle, slightly swollen in the middle. The outer surface is smooth. The circle is

41

composed of a dry, brittle, whitish substance, beneath which are several layers of minute, yellowish white eggs pressed closely together side by side. If you have the patience to count the eggs in a single disc you will find two or three thousand of them.

Should you be so fortunate as to watch the hatching of these eggs some moonlit summer's night, you would see myriads of curious little creatures burst from the egg mass and drop to the water below. Each one is nearly a fifth of an inch long, and has a large head, six long legs, and a pair of

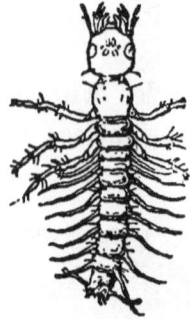

FIG. 41. — Young Dobson, much enlarged.

thread-like projections on each ring of the body (Fig. 41). As soon as they reach the water they seek the shelter of stones and pebbles. Soon they begin to search for little water worms of almost any kind weaker than themselves. This search they keep up as long as they live in the water.

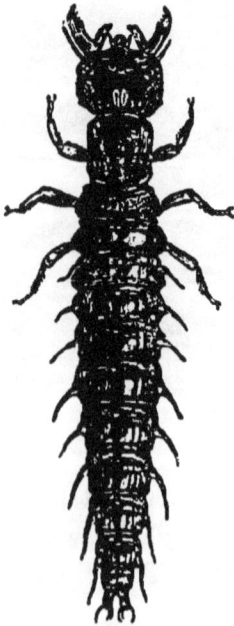

FIG. 42. — Full-grown Dobson, natural size.

These little creatures are young "dobsons." Even in this early stage they must prove formidable foes to the soft bodies of the young May flies, as well as of the larvæ of stone flies, dragon flies, and the various other water-loving insects upon which the dobsons feed. As the days pass they grow gradually in size, and are able to prey upon larger and larger insects to satisfy their increasing appetites.

At long intervals the dobsons moult. Apparently they only shed their skins about six times during the three years of their existence in the water. Except for their increase in size there is little change in their general appearance during this period.

I am sure many boys will recognize the picture of the full-grown dobson larva shown in Fig. 42. These larvæ are also often called "crawlers" by fishermen. They are frequently used for bait in fishing. As will be seen, the head is large and broad, and is provided with a pair of formidable jaws. There are six long, strong legs on the three rings of the body just behind the head. On each of the hinder rings of the body is a pair of tube-like threads, by means of which the insect breathes the air in the water.

The dobsons prefer to live in the rapids of rivers and creeks, where the water courses swiftly over the rocks. They may often be caught by disturbing loose stones in such streams, having some one else hold a net below so that the dobsons will be carried into it by the rapid current.

THE DOBSON BECOMES A HELLGRAMITE.

IN the spring or early summer of the third season of its existence the dobson leaves the stream in which heretofore it has developed, crawls up the bank and seeks the shelter of some log or stone lying above the water level. In the soil beneath this it hollows out an oval cell, in which it generally lies a week or more before changing to the pupa state. Finally it transforms to the pupa (Fig. 43).

FIG. 43. — Pupa of Dobson.

Compared with its previous form, the most notable change that has taken place in the dobson is due to the absence of the threads along the sides of the body and the presence of the wing pads upon its back. In these the enormous wings of the adult insect are to be developed.

The dobson remains in this pupa state two weeks. Then it changes again and the adult hellgramite appears. This also is a curious-looking creature. It is of a dull gray color. The head is large and furnished with prominent eyes and long feelers or *antennæ*. The legs are long and stout, and the wings of immense size. They are membranous with a network of veins. When at rest they project some distance beyond the end of the body. The males have a pair of very large pincher-like jaws (Fig. 44).

These adult hellgramites are nocturnal insects. They remain at rest during the day in such shelter as they can find, and fly about at night. Occasionally one will come into a lighted room through an open window. Sometimes

44

they may be seen in the evening flying about electric street lamps.

The female hellgramites deposit their eggs in the circular masses upon the sides of rocks in running streams. They select such situations that the little dobsons when hatching can drop directly into the water. Some such egg masses deposited on a rock in the middle of a New England river are shown in the photograph reproduced in Fig. 40.

In regions like the Mississippi Valley where rocks are less abundant than in the eastern states these egg masses are deposited upon the large leaves of sycamore and other trees overhanging the water.

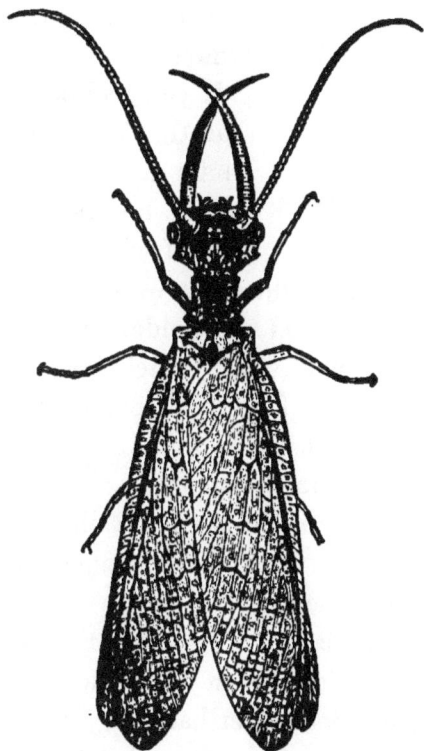

Fig. 44. — Hellgramite.   Male, natural size.

THE APHIS LION.

DID you ever see some little threads with knobs on their tips sticking up from the surface of a twig or leaf? They look like Fig. 45, and may readily be found during summer upon a great variety of plants.

FIG. 45. — Eggs, enlarged.

If you look at the threads closely, you will find that the knobs are white and nearly egg-shaped. Under a hand lens you will see that the egg-shaped bodies differ from the threads upon the ends of which they are fastened.

These white knobs are the eggs of a pretty four-winged insect called the golden-eye, or lace-winged fly. The threads are simply stalks which hold them up above the surface out of the reach of ladybird beetles and other insects that eat such eggs.

Should you be able to watch one of these groups of knobbed threads for some time, you would be likely to see a curious little larva hatch from each of the eggs. The time of hatching is likely to vary, some emerging from the eggs much earlier than others.

The larvæ which have thus been cradled in the air are called aphis lions, from their habit of feeding upon aphides or plant lice. Soon after hatching, they wander over the plant in search of prey. Almost any small insect will answer for this purpose, but plant lice form the principal item in the daily bill of fare. Each aphis lion has a large pair of jaws projecting forward from the head. These jaws are so constructed that each is a hollow sucking tube, through which the life-blood of the victim is drawn, as well as an organ for seizing and piercing the prey.

Like other larvæ, the aphis lions cast their skins occasionally

FIG. 46. — Aphis Lion eating Insect, enlarged.

as they increase in size. They become fully developed in about two weeks. They are then nearly an inch long, and of the form shown in Figs. 46 and 47. The different species vary considerably in color, but most of the aphis lions are of mottled shades of blue, brown, black, and white.

Should you examine under a lens the long, sucking jaws of the aphis lion, you would find that each jaw consists of two parts, each with a groove running along the inside, so that when the parts are fitted together the tube is formed.

The full-grown aphis lion prepares for the change to the pupa state by rolling itself together compactly. Then it spins from the posterior end of its body a spherical silken cocoon, so small that one must wonder how so large a larva stays inside of it. The completed cocoon

FIG. 47. — Back View of Aphis Lion, enlarged.

FIG. 48. — Cocoon, enlarged.

is shown in Fig. 48. It is about the size of a small, smooth pea, and is of a pearly white color, generally mottled in places with black.

Within this tiny silken ball the larva becomes a pupa. A short time afterwards the pupa gnaws a circular cap out of the cocoon and escapes into the outer world. Then it changes into an adult lace-winged fly.

THE GOLDEN-EYE, OR LACE-WINGED FLY.

THE insect which comes out of the pupa is very different from the larva that went into it. It is a delicate-looking creature a little over half an inch long, of a pale green or bluish green color, with beautiful golden eyes standing out prominently on the sides of the head. From the head also project two long and slender feelers or antennæ. Under a lens these are seen to be clothed with many fine hairs (Fig. 49).

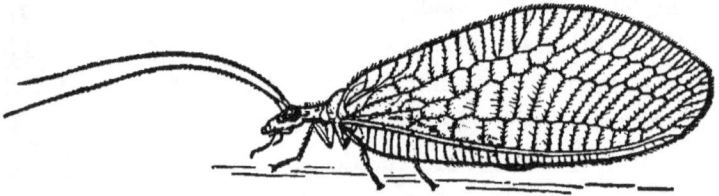

FIG. 49. — Lace-winged Fly, enlarged.

The part of the body directly behind the head — called the *pro-thorax* — is wide and flattened. It bears a single pair of legs. The next part of the body — the rest of the *thorax* — bears two pairs of wings and two pairs of legs.

The legs are long and slender, and of much the same color as the body. At the tip of each foot are two recurved claws.

The wings are very large in proportion to the size of the body. They are composed of a thin, transparent membrane, stretched between a beautiful network of delicate, greenish veins, which bear rows of brownish hairs. The front and hind wings are very similiar in shape, the hind ones being somewhat smaller.

48

When the lace-wing is at rest the wings are folded in a nearly vertical position, so that they project beyond the hind end of the body.

The female lace-wings deposit their eggs on the tips of long stalks. The stalk is drawn out from a liquid secretion, which hardens on exposure to the air, and the egg is then glued upon the tip. By thus placing the eggs up above the leaf surface the insect saves them from being eaten by ladybird beetles and other insects. A week or more after they are deposited the eggs hatch into young aphis lions which, like their namesakes of the desert, go about seeking what they may devour.

While the beauty of the color and structure of the lace-winged fly appeals strongly to the eye of the nature lover, the insect has a very different effect upon his nose, for these delicate creatures emit probably the most disagreeable odor of any insect. It is worse, to many minds at least, than that given off by the bedbug and its allies, or even the noisome pestilence of the carrion beetles. How so small an insect reared from infancy upon a cleanly diet of the juices of freshly killed insects, spending its resting period in a " glistening white cocoon which looks like a large seed-pearl," and feeding little, if at all, in its adult state, can develop so disagreeable a stench is indeed a wonder. The purpose of the odor is doubtless to protect the lace-wings from the attacks of birds and other enemies.

DURING June few insects are more abundant than the aphides or plant lice. These are small, green, brown, or black insects that may be found crowded on the young twigs of apple, cherry, and other trees. They are very similar to the little " green fly " so often injurious to house plants.

By a little searching you may often see in the midst of these colonies of plant lice on the branches of trees and shrubs a few small white eggs, much longer than wide. Sometimes the surface of these eggs is smooth, but often the eggs are covered with distinct raised lines.

These are the eggs of the bright-colored, two-winged flies called syrphid flies. A few days after they are laid the eggs hatch into little footless maggots, somewhat the shape of a very long triangle.

At the head end, each of these maggots is furnished with a pointed beak. By means of this beak the young larva pierces the body of the nearest plant louse. It then generally holds its victim up in the air while it sucks the life-blood.

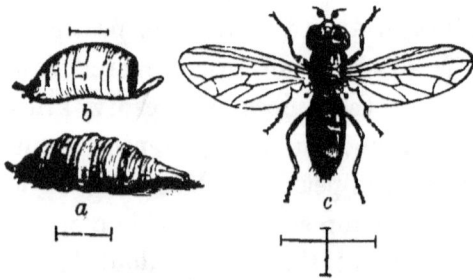

FIG. 50. — Syrphid Fly: *a*, Larva; *b*, Pupa Case; *c*, Fly. Straight Lines show the Natural Size.

The plant lice are small, and a single one does not long satisfy the appetite of the syrphid larva. As soon as the first victim is sucked dry its shriveled skin is cast aside and the fatal dagger is thrust into the side of another.

Thus day after day the slaughter continues, the syrphid moving slowly onward as it disposes of the aphides immediately before it, and leaving in its path the shrunken remains of its unfortunate victims, scores of which are required for each day's subsistence.

The syrphid larva casts its own skin occasionally as it increases slowly in size, attaining when full-grown a length of from one-half to three-quarters of an inch. At this time the larvæ are of the general form represented in Fig. 50, a, and are of various mottled shades of brown, yellow, red, and white.

The larva attains its full growth a few weeks after hatching from the egg. It is then ready to change to the pupa state. To accomplish this the lower surface of its body becomes firmly attached to the twig or leaf upon which it is resting; the outer skin then hardens and turns brown, while inside the larva shrinks away from it and becomes a pupa. The hardened outer skin which encloses and protects the pupa is called the *puparium*. The insect remains in this quiet condition ten days or more; then the pupa changes to a two-winged fly, which pushes open a lid at the larger end of the puparium and escapes. Such an empty puparium is represented in Fig. 50, b. The adult fly is a sunshine-loving creature that flies very rapidly, and freely visits flowers in search of nectar and pollen.

THE FLOWER SPIDER.

During summer throughout most of the United States one can scarcely study flowers in the field an hour or two without finding one or more blossoms in which there is a small crab-shaped spider, of a white or yellow color, resting upon the petals or concealed within other parts of the flower. Often, though by no means always, the white spider will be found upon the white blossoms and the yellow spider upon the yellow blossoms, so that the color resemblance renders the flat expanded body inconspicuous. As the season passes the flowers change, and these spiders choose new blossoms for their abodes. In early summer you may find them upon the buttercups and daisies, while later in the season, thistles, sunflowers, and other large blossoms are favorite habitations.

Not infrequently you may see a butterfly resting quietly upon a flower. You approach cautiously, thinking to grasp it in your hand before it awakes from its nap. Suddenly you catch it between thumb and finger and hold it up to find one of these spiders attached to its body. There is the secret of the easy capture : the butterfly is dead and was furnishing a meal to the spider. There also is the explanation of the spiders' presence on the flowers : they are lying in wait for the butterflies and other insects that come to sip the nectar or nibble the pollen of the blossoms. They lie motionless for hours, until the unwary victim comes within reach ; then there is a sudden spring and the insect is captured. The spider sucks the juices of the body and drops the juiceless fragments to the ground. A great variety of insects are captured in this way, flies and butterflies seeming to be the most numerous victims.

These spiders belong to the family called the crab spiders, because of their resemblance to a crab and their habit of running sideways rather than straight ahead. The hind part of the body (*abdomen*) is large and flat, while the front portion is small; the eight legs attached to the latter extend sideways for some distance.

FIG. 51. — Flower Spider and its Victim.

An interesting study of the relation of color to habits of life and surroundings is presented by this family of crab spiders. There are two principal groups having different habits as regards the places in which they choose to live. The flower-loving group are white or yellow, sometimes with brilliant spots or streaks that help their resemblance to parts of a flower. Most of the other members of the family live upon the bark of trees, logs, shrubs, and in similar situations, and are colored in various shades of gray and brown. This resemblance to their surroundings is useful to these spiders in two ways: (1) by concealing them from birds or other enemies; naturalists call this *protective resemblance;* (2) by concealing them from the insects which they catch for food; this is called *aggressive resemblance.*

www.ingramcontent.com/pod-product-compliance
Lightning Source LLC
Chambersburg PA
CBHW022014190326
41519CB00010B/1512